故園畫憶

庚寅中秋

韩启德题

故园画忆系列

Memory of the Old Home in Sketches

老舟山风情

The Customs of Old Zhoushan

罗　枫　绘画 撰文

Sketches & Notes by Luo Feng

學苑出版社

Academy Press

图书在版编目（CIP）数据

老舟山风情 / 罗枫绘画、撰文. — 北京：学苑出版社，2019.8
（故园画忆系列）
ISBN 978-7-5077-5789-7

Ⅰ. ①老… Ⅱ. ①罗… Ⅲ. ①建筑画—作品集—中国—现代②舟山—概况
Ⅳ. ①TU204.132②K925.53

中国版本图书馆CIP数据核字（2019）第179617号

出 版 人：孟　白
责任编辑：杨　雷　张　翔
编　　辑：陈柯宇
出版发行：学苑出版社
社　　址：北京市丰台区南方庄2号院1号楼
邮政编码：100079
网　　址：www.book001.com
电子信箱：xueyuanpress@163.com
联系电话：010-67601101（营销部）、010-67603091（总编室）
经　　销：全国新华书店
印 刷 厂：河北赛文印刷有限公司
开本尺寸：889×1194　1/24
印　　张：4.5
字　　数：120千字
图　　幅：92幅
版　　次：2019年9月北京第1版
印　　次：2019年9月北京第1次印刷
定　　价：38.00元

目　录

自然景观

生活民俗

Contents

Landscape

Folk Life

自　序

历史的变化纷繁复杂，有时甚至惊天动地、天翻地覆。几百年几千年，甚至几万年过去了，地域的变化、环境的改变、朝代的更替、人类的迁徙，历史刻录下来不同的文化信息，就成为我们探究历史的生动影像！

岛因海而生、因水而兴。舟山群岛是我国最大的群岛，唐开元三十六年（738年）设县城，是我国唯一的以群岛设立而成的县城。舟山群岛中的普陀山与五台山、峨眉山、九华山并称为中国佛教四大名山。佛教历史源远流长，唐咸通四年（863年），日僧慧锷从五台山请进一尊观音圣像乘船回国，行至普陀莲花洋受阻，便登潮音洞岸边，留观音像供奉于渔家，后世称“不肯去观音院”，普陀山遂成观音菩萨应化道场。自唐以来，有20多位帝王为求国泰民安亲临或遣内侍来此求拜观音，或赐金银修寺建院。历史上，倭寇袭扰，南明抗清，鸦片战争和数次海禁，充满了波折与磨难。每次灾难过后，寄托着舟山人民希望与梦想的新家园又会在海上重新崛起。

历史雕刻了时光，也留下了舟山的性格和特色。老街巷的地名，记录着城市的前世今生、市井百态和人文血脉。回顾这一幕幕历史的画卷，我希望透过这一张张钢笔画，再现这片土地上千百年来的历史风貌。寻找蕴藏在海岛上的人文古迹，让我们一起在这些丝丝缕缕、密杂繁纷的线条中，领略老舟山的风味吧！

在画册即将出版的阶段，我想再次衷心地感谢在《老舟山风情》的编创过程中为我提供帮助的热心人士们，陈士定、王建富以及舟山文广局的工作人员，在他们的帮助与支持下我得以更好、更完整地展现舟山的历史与风情。

罗　枫

2019年3月29日

Preface

There have been numerous, complicated and sometimes earth-shattering changes in history over the last hundreds and thousands of years. These changes include regional, environmental, vicissitude of dynasties and human migration. Thanks to historical records, our exploration of history has left us with vivid images! Islands live and flourish in the seas bring life and nourishment to the waters. The Zhoushan archipelago is a group of more than 400 islands off the northern coast of Zhejiang province in eastern China and is considered as the largest archipelago in China. Putuo Mountain, Wutai Mountain, Emei Mountain and Jiuhua Mountain are widely known as the four famous mountains of Chinese Buddhism. They have a long history in the region, which can be dated back to the fourth year of Xiantong in the Tang Dynasty (863), when a Japanese monk Hui E, who was carrying a statue of Avalokitesvara Bodhisatttva from Wutai mountain to the motherland by boat, and stuck on Putuo mountain in the Lotus Ocean. He stepped onto the shore at Chaoyin cave and left the statue of Avalokitesvara Bodh, and dedicated it to fishermen on the mountain, as it was said that the Avalokitesvara Bodh loved the mountain so much that she refused to leave the mountain. Since the Tang Dynasty, more than twenty emperors have paid visits to the temples on the mountain, or sent their eunuchs to worship the Guanyin Bodhisattva and donated money to build monasteries in the hope of bringing peace and prosperity to the country. However, history has not been so kind, it has thrown many twists and turns in our direction. For example, the invasion and harassment from Japanese pirates, the movement of the Southern Ming Dynasty resisting the Qing armies, the Opium wars and several maritime prohibitions. However, every time after disasters happened, Zhoushan people would rise again on the sea, filled with hope and dream to build their new homeland.

Although historical events have carved many changes into the area, it has also managed to leave and preserve many of the old characteristics of Zhoushan.The names of the old streets and lanes record the past and present life of the city including various lifestyles and human cultures. After carefully examining scrolls of historical panorama, I will try to present you a vivid image of the cultural heritage that has been hidden on these islands for thousands of years, also reproduce some of the historic characters once again by studying antique pen drawings. Let's sit back and enjoy the flavor of old Zhoushan Mountain in artistic wisps of intricate lines.

I would like to sincerely thank Chen Shiding and Wang Jianfu, the editors of *The Customs of Old Zhoushan*, also a big thank to the enthusiasts of the Zhoushan Broadcasting and Cultural Bureau. With their help and support, I am able to display the history and customs of Zhoushan as best I can.

Luo Feng

2019.3.29

历史遗存
Historical Preservation

祖印寺

位于定海区城关昌国路 98 号，始建于五代后晋天福五年（940 年），迄今已有千余年历史。现存寺院在中轴线上依次排列的有山门、天王殿、大雄宝殿、后大殿，在中轴线左侧有鼓楼、厢房、斋堂等，在中轴线右侧有厢房、钟楼。全寺建筑面积为 2212 平方米，占地 5125 平方米。除正三间系近年重建外，其余皆为清代建筑。

Zuyin Temple

The temple is located at No. 98 Changguo Road, Chengguan Area, Dinghai District. It was bulit in the 5th year of Tianfu (940) of the Later Jin Dynasty. The building area of the temple is 2212 square meters and the entire grounds covers an area of 5125 square meters. Apart from the recent reconstruction of three compartments, the others buildings are all from the Qing Dynasty.

长寿禅院

位于岱山县秀山乡黄泥坎村。始建于后汉乾祐二年（949 年），明代海禁期间院舍遭荒废，清嘉庆七年（1802 年）重建。长寿禅院文化底蕴深厚，院中藏有赵朴初、沙孟海、俞德明、妙善等名人墨宝。

Longevity Monastery

The temple is located in the Huangnikan village in Xiushan Township, Daishan County. The monastery was founded in the second year of Qianyou of the Later Han Dynasty (949). The courtyard was abandoned during the sea ban period in Ming Dynasty, and was then rebuilt in the 7th year of Jiaqing of the Qing Dynasty(1328). Longevity Monastery contains many precious pieces of calligraphy and paintings of celebrities.

多宝塔

位于普陀区普陀山镇普济路 253 号。始建于元元统二年（1334 年），该塔系宣让王帖木儿不花“施钞千锭”，命高僧孚中怀信禅师购太湖石建造，故又名“太子塔”。塔高 32 米，四面五层，有台无檐，塔下基座高大，围绕栏杆，雕刻佛、菩萨和供养人像，并饰螭首，形象生动。塔顶做宝箧印经塔的形状，四角出山花蕉叶，塔刹呈仰莲宝瓶。多宝塔为典型的元代建筑工艺，被称为普陀山三宝之一。

Multi-treasure Pagoda

It's located at NO.253 Puji Road, Putuo Mountain Town, Putuo District. It was founded in the 2nd year of Yuantong of the Yuan Dynasty (1334). The tower stands at 32 meters high, with five floors that all have verandas but no eaves under which is a huge pedestal. There sit the statues of Buddha, Bodhisattva and their providers are carved around the balustrade, with the decoration of the heads of legendary dragons, vivid and graphical. Mufti-treasure Pagoda is one of the three treasures of Putuo Mountain which is a typical architecture of the Yuan Dynasty.

梅福禅院

位于普陀区普陀山的梅岑山脉西巅。明万历年间普陀寺住持如迥创建，又称“梅福庵”“梅岑禅院”。清光绪元年（1875 年），扩建殿阁，遂成现在规模。庵院临坡而起，依岭而筑。虽规模不广，但其内整洁无染，梵宇清静；其外林木幽深，潮音遥闻。今为尼师庵院。

Meifu Monastery

Located in the western foothills of the Meicen Mountain Range in Putuo Mountain, Putuo District. During the Wanli Period of the Ming Dynasty, the abbot of Putuo Temple was created, this gave the temple the name “Mei Fu An”, and later changed to “Mei Cen Monastery”. In the first year of Guangxu of the Qing Dynasty (1875), the temple was expanded, and the scale was changed to its current size, and at the time was renamed as “Meifu Monastery”.

万峰禅院

位于定海区东海西路 69 号海山公园内的龙峰山上，又名万峰庵。该禅院创建于清嘉庆年间，在禅院山门的墙上镌刻有《万峰禅院志》，该志讲述了禅院创建的来龙去脉。禅院供奉的是日光菩萨和月光菩萨，以日光菩萨圣诞为香期。

Wanfeng Monastery

Also known as Wanfeng Nunnery, situated on the Longfeng Mountain in the Haishan Park, No.69 Donghai West Road, Dinghai Distict. The Monastery was founded in the reign of Jiaqing of the Qing Dynasty. The wall of the gate to the Monastery is engraved with *The Annals of Wangfeng Monastery,* which tells of the creation of the monastery. Wanfeng Monastery worships the Sunshine Bodhisattva and the Moonlight Bodhisattva, also the divine birthday of the Sunshine Bodhisattva by burning Christmas incense.

慧济寺

位于普陀区普陀山佛顶山上，为普陀山三大寺之一。始建于明代，清乾隆五十八年（1793 年）扩庵为寺，清光绪三十三年（1907 年）经朝廷批准请得《大藏经》及仪仗，钦赐景蓝龙钵、御制玉印等。全寺建筑别具一格，依山就势，横向排列，殿堂宽敞壮丽，整座寺院深藏于森林之中，以幽静称绝。

Huiji Temple

The temple is located on the Foding Mountain, of Putuo Mountain, in Putuo District, and is one of the three major temples of Putuo Mountain. It was built in the Ming Dynasty, and added to in the fifty-eight years of Qianlong in the Qing Dynasty (1793). It was expanded again in the years of Guangxu in the Qing Dynasty. The whole temple building has a unique style of its own being arranged in the horizontal direction. The temple is spacious and magnificent, and the whole temple is weighed in quietness as it is completely hidden deep within a forest.

奎光阁

位于定海区鳌山附近。始建于清道光十六年（1836 年），不久毁于台风。清同治十年（1871 年）重建。为六边形三层塔状建筑，每一层均飞檐翘角，状如大鹏展翅。重檐攒尖顶，当顶倒覆一涂釉瓷缸，上置深色琉璃宝葫芦，阁底层朝北开一道正门，二、三层墙上四面开窗，整座塔古朴无华、典雅凝重。现塔已无存。

Kuiguang Tower

It was located near the Aoshan Mountain in Dinghai District. It was built in the 16th year of Daoguang of the Qing Dynasty (1836) and was soon destroyed by a typhoon. It was then reconstructed in the tenth year of Tongzhi of the Qing Dynasty (1871). The whole tower is quaint and elegant in the drawings. However, the tower has now disappeared.

传灯庵

位于岱山县长涂镇长涂港北口海面上。始建于清光绪二十四年（1898 年），为邑人金品三之子金霖募捐，是为了在西鹤嘴上点灯为过往船只照明导航而建。该庵有前后大殿各三间，南北厢房各七间，以及山门、围墙等。古朴典雅，是一所保存完好的古建筑物。

Chuandeng Nunnery

It is located on the sea surface of the north port of Changtu Town, Daishan County. It was built in the twenty-fourth year of Guangxu of the Qing Dynasty (1898). It was built in order to shed light on the Xihe Mouth for clear navigation of passing ships. It is a simple and elegant yet well-preserved ancient building.

白云宫

位于普陀区东极镇东极岛东福山上，始建于清宣统二年（1910 年），白云宫有五个大殿，分别设了宫、殿、堂、观、洞，供奉了释、道、儒、神、怪等不同的宗教对象，让人们各取所需。但是公众最信奉的却是白云娘娘，所以又称“白云宫”。白云宫内无道无僧，全凭村子里的善男信女自行打扫拭拂。

White Cloud Palace

Located in Dongfu Mountain, Dongji Island, Dongji Town, Putuo District, it was built in the second year of Xuantong of the Qing Dynasty (1910). Different religious objects such as Tao, Confucianism, God, and so on allow people to say their own prayers. But the local public worship the Baiyun Empress, so it is also called the White Cloud Palace.

石礁庙

位于定海区石礁社区。石礁乡《千岛乡情》载："桥头施在宋前还是一个避风港湾，港名曰'方潭洋'，洋中有一礁石，一次一艘商船行至方潭洋面，不幸触礁。由于礁石离岸较近，船上 10 余人脱险被救，为感谢苍天再生之恩，便在礁石上建造一庙，名曰'石礁庙'，供奉香火，石礁因此得名。"

Reef Temple

Located in the Shijiao community of Dinghai District. Before the Song Dynasty, the port of Fang Tanyang was used as a safe haven from a reef in the ocean. Once a merchant ship travelled across Fangtan Ocean. Unfortunately, his boat hit the rocks. As it was a part of the reef closer to the shore, more than 10 people on board were rescued from danger. In order to thank the heavens for their reincarnation, a temple was built on the reef, named Reef Temple.

唐脑山灯塔

位于嵊泗县洋山镇圣港社区唐脑山（岛）西顶。始建于清光绪三十四年（1908 年），由英国人建造。灯塔为白色砖房，高 7 米，灯高海拔 30 米，建筑面积 150 平方米，现归上海航道局管理，目前为无人灯塔。是我国南部沿海航线出入长江口的重要导航设施。

Beacon of Tangnao Mountain

It is located in the west of Tangnao Mountain (Island), Shenggang Community, Yangshan Town, Shengsi County. It was built by British in the thirty fourth year of Guangxu of the Qing Dynasty (1908). The Beacon is made of white bricks. It is now managed by the Shanghai Waterway Bureau and is currently an unmanned Beacon. It is very important to the southern coastal navigational route of the Yangtze estuary.

继思桥

位于定海区双桥镇南山社区南山后房6号民居门前。始建于乾隆五年（1740年），桥体为钓山版的“糯米红”石材砌成，桥栏上是又长又厚的红石条，红石板也已经显得青痕斑驳，桥北向镌刻有“乾隆庚申造”字样，南向镌刻“继思桥”三个行书大字，左右镌刻有“嘉庆柒年陆宗重建”字样。

Jisi Bridge

It is located in Nanshan Community, Shuangqiao Town, Dinghai District. It was built in the 5th year of Qianlong of the Qing Dynasty (1740). The body of the bridge is made of stone , while the bridge railings are long and thick with red stone. The north side of the bridge is engraved with the words “Qian Long Geng Shen Zao”, and the words”Ji Si Qiao” engraved in the south. It also has the words “Jia Qing Qi Nian Lu Zong Chong Jian” engraved in the left and right.

寺岭桥

位于岱山县小沙镇庙桥社区（寺岭村），是原来小沙乡民进定海县城的必经之地。桥面长约 13 米，宽约 4 米，石桥拱门宽 3 米，高 6 米。此桥为舟山现存最古老的原生态石拱桥，确切建造年代目前难以考证，但据附近古驿站设立的年份为清嘉庆年间来看，此桥的修建应当在此之前。

Siling Bridge

It is located in Miaoqiao Community (Siling Village), Xiaosha Town, Daishan County. It was the only way for the villagers from Xiaosha Town to enter Dinghai County. This bridge is the oldest surviving stone arch bridge in Zhoushan.

娘娘桥

位于普陀区展茅街道茅洋社区松山村，是一座乱石拱桥。始建于清乾隆年间，距今已有 200 多年的历史。相传古时迎亲，新娘子花轿需过此桥进村，不得他道而行，故将此桥取名娘娘桥。该桥跨越茅洋溪，南北走向，桥面长 9.7 米，宽 2.45 米，拱高 3.3 米，拱跨 3.6 米。桥面呈流线型，古藤缠绕，十分美观。

Bride Bridge

Located in Songshan Village, Maoyang Community, Zhanmao Street, Putuo District. This rocky, arched bridge was built in the in the reign of Qianlong in Qing Dynasty. According to legend, in ancient times, the bride in the sedan chair needed to kiss the bridge whilst crossing into the village. Therefore, the bridge was named the Bride Bridge. The bridge deck is streamlined with ancient beautiful entangled vines.

寺岭石拱桥

位于定海区兴华社区寺岭村西侧，始建于清嘉庆年间。桥呈南北走向，架于龙潭坑上游，利用拱力作用，依岩而垒，精心砌筑。

Siling Stone Arch Bridge

It is located on the west side of Siling Village, Xinghua Community, Dinghai District. The bridge was built in the jiaqing period of the Qing Dynasty. The bridge was erected on the upper reaches of Longtan pit, and built in the shape of an arch with local rocks.

甩龙桥

位于定海区小沙镇华厅村上游。始建于清道光二十四年（1844 年），重修于同治年间。桥的主体为单孔石拱桥，桥身长 7.7 米，桥面宽 2.36 米，净跨 3.7 米，桥身以东北－西南走向横跨大溪坑，连接华厅村和新光村，是当地的重要交通通道，至今仍在使用。

Jilt Dragon Bridge

Located in the upper reaches of Huating Village, Xiaosha Town, Dinghai District. It was built in the 24th year of Daoguang in the Qing Dynasty (1844), and was rebuilt in the reign of Tongzhi. The bridge reaches northeast-southwest and runs across Daxi pit, connecting Huafang Village and Xinguang Village. It is an important channel in the area and is still in use today.

环龙桥

位于岱山县高亭镇黄泥岙村水库西侧，为清代所建单孔石拱桥，拱券用不规则石块并列砌建，位于水库中，平时在水底下，水位低时才露出水面来。

Ring Dragon Bridge

It is located in the west side of the Huangni'ao Village Reservoir in Gaoting Town, Daishan County. It is a single-hole stone arch bridge built in the Qing Dynasty. The arch vouchers are built in parallel to each other with irregular stones that stand on the bottom of the reservoir. Usually the bridge is under water until the water level is very low, then the true beauty of the bridge is exposed.

金井桥

位于定海区金塘镇柳行街上下界分界处，原名“望仙桥”，因能在桥上遥望仙人山而得名，河对面建有“金井庙”。金井桥横跨金塘溪，呈东南－西北走向。桥面由长方形石条铺成，北侧为阶梯。桥面宽2.1米，长10米，栏板两侧均刻有“金井桥”三个字，每侧各有五个望柱。美观坚固，古意十足。

Golden Well Bridge

Located at the joining of the lower boundary of Liuxing Street, Jintang Town, Dinghai District. Golden Well Bridge spans the Jintang River and runs southeast-northwest. The bridge deck is 2.1 meters wide and 10 meters long. The words "Jin Jing Qiao" are engraved on both sides of the slab, and each side has five columns. The bridge is beautiful and sturdy and full of ancient meaning.

定海古城·南城门

千余年来，定海早在唐代便设县造墙，唐开元二十六年（738 年），舟山始设县，名“翁山”，县治移建于镇鳌山下，建城墙周广五里。宋熙宁六年（1073 年），置昌国县，城墙增至九里。明永乐十五年（1417 年），加修城墙周围七里。清康熙二十九年（1670 年），置定海县，重修定海城墙。图为依据乾隆五十八年（1793 年）英国画家亚历山大所绘“定海南城门”改绘而成。

Dinghai Ancient City, South Gate

In twenty-sixth year of Kaiyuan of Tang Dynasty (738), Zhoushan was founded as a county, named "Weng Shan". After thousands of year of expansion, Dinghai City Wall has formed into what we see today. The illustration is based on the painting of "Dinghainan City Gate" by British painter Alexander in the fifty-eighth year of Qianlong of Qing Dynasty (1793).

定海古城·书院弄

位于定海区，舟山群岛于康熙二十三年（1684 年）开海禁，设定海县，当时百废待兴，教育断层，江苏人缪燧任县令后，自捐俸银办学，其后响应甚众，逐成书院弄。

Dinghai Ancient City, Academy Lane

Located in Dinghai District, the Zhoushan Islands were released from the maritime prohibition and opened in the 23rd year of Kangxi of Qing Dynasty (1684), who set up Dinghai County. After Miao Sui from Jiangsu province was appointed as the county magistrate, he donated his salary to run a school, which was popular and Later became an Academy Lane.

定海古城·御书楼

位于定海区昌国路103号定海一中院内，始建于清康熙二十八年（1689年），御书楼是为了专门敬奉康熙皇帝的亲题匾额“定海山”而建。后世多有修葺，整座建筑坐北朝南，占地面积264平方米，建筑面积128.8平方米。设有台门、围墙、正楼等建筑。

Dinghai Ancient City, Royal Calligraphy Pavilion

It is located in Dinghai No. 1 Middle School Courtyard, No.103 Changguo Road, Dinghai District, it was built in the 28th year of Emperor Kangxi of Qing Dynasty (1689). The Royal Calligraphy Pavilion is dedicated to worship Emperor Kangxi. The whole complex including the main buildings, walls and gates is facing south and it has all been well-preserved.

定海古城·封火墙

封火墙在当地也称“城门”，始建于光绪十七年（1891年），位于定海当时最热闹的三条街，中大街、西大街、东大街上。这三条街上店铺林立，人口众多，店家紧挨着连片的民居，一旦失火损失惨重。当时的定海知县为了避免火灾带来的严重损失，动员民众集资修建了八道封火墙，又称“公墙”，并立碑文。

Dinghai Ancient City, Fire Seal

The Fire Seal, also known as the “city gate” was built in the seventeenth year of Guangxu of Qing Dynasty (1891). At that time, the West Street, Middle Street. East Street were very busy with the hustle and bustle of the people, but this caused many fires and the losses were heavy. Therefore, the magistrate of Dinghai County mobilized the people to raise funds to build eight fire walls.

定海古城·街景（一）

定海是一座历史悠久、古迹众多的的千年古城，古城内保存有明清时期的街区，其中的西大街、中大街是晚清商业街市，街宽四五米，在当时来讲是真正的商业大街。街边两旁排列着一家家的店铺，店铺一般为木结构上下二层、配封火墙，二层的檐廊是当年俯瞰街市繁华景色的最佳场所。

Dinghai Ancient City, Street View (1)

Dinghai ancient city preserves the streets of the Ming and Qing Dynasties. The West Street and the Middle Street are both the late Qing Dynasty commercial market. The streets are 4-5 metres wide and have shops standing on both sides.It was a true commercial street at that time.

定海古城·街景（二）

定海由于环境和历史原因，传统民居建筑呈现多样化的格局。明清时期长期海禁，渔业、农业成为当时的两大主要行业，此时的民居呈现出农业社会格局下的海岛民居建筑特色，行业特点的差异导致了居住方式的差异。明清以后，资本主义生产方式开始在中国萌芽，商贸、运输开始成为当地新兴的营生方式，商贾宅院从此成为海岛民居建筑的一个新的亮点。

Dinghai Ancient City, Street View (2)

Dinghai traditional residential architecture styles are diversified, including not only the characteristics of the residential form of the agricultural society during the Ming and Qing dynasties, but also the merchant houses after the emergence of capitalist production methods after the Ming and Qing Dynasties.

定海古城·街景（三）

定海古城虽小，当年却十分繁华，各类店铺林立，定时定期的商业街市、集市不断。老街点点的石板路，弯弯曲的小型堂，记录着定海当年的繁华和荣光。图为当年定海街巷叫卖的小贩。

Dinghai Ancient City, Street View (3)

Although the ancient city of Dinghai was small, it was very prosperous with many different shops, regular commercial markets and fairs. Stone slabs ditted in the old street. sinuate small halls record the prosperity and glory of Dinghai in those years. The picture shows a hawker selling his goods in Dinghai's street.

东沙镇

位于岱山县，东沙镇建制于唐，兴盛于清。得渔、盐之利，在清末民初，商贸就得到飞速的发展。岱衢洋盛产岱衢族大黄鱼，每年渔汛季节，沿海山东、江苏、浙江、福建等地渔船云集，“船过数千，人过数万”。与鱼货交易伴随的是一批“老字号”商铺的兴起，诸如严永顺米店、三阳泰南货、鼎和园香干、聚泰祥布庄、王茂兴老酒等。

Dongsha Town

Located in Daishan County, Dongsha Town was established in the Tang Dynasty and flourished in the Qing Dynasty and continued into the Republic of China, trade and commerce rapidly developed. Every year, the annual fishing season would bring thousands of boats, carrying tens of thousands of people from all over China. The fish trade aided the rise of some time-honoured shops.

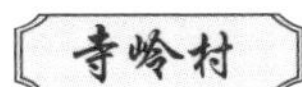

位于定海区小沙镇，是定海海拔最高的古村落，寺岭村因旁有春祥寺而得名，已有几百年的历史。古村落已繁衍数代，现全村搬迁下山，只有枯树老藤旧屋依傍山间，但仍能让人回忆起当年村舍的热闹场景。

Siling Village

Located in Xiaosha Town, Dinghai District, is an ancient village on the highest altitude in Dinghai. It has a history of several hundred years. Siling Village is named after Chunxiang Temple. Now the whole village has moved down the mountain, and only the old houses of dead trees and old vines remain nestled in the mountains, but still act as a reminder of the lively scenes that once appeared in the village.

桑园弄

位于定海区，因当地居民种桑而得名，是旧时定海城内难得的一块林木成片的区域。这里既有清朝末年的传统民居又有民国初年的传统民居，当年附近还有长生桥、永福桥、昭忠祠等建筑。

Mulberry Garden Lane

Located in Dinghai District, it is a rare forest area in old Dinghai City. It is named after the local residents who planted and grew mulberry bushes.

皇帝堂

位于定海区昌国街道留方社区 18 号。是当地王姓人家的祖宅，约有 300 多年历史，相传乾隆下江南时曾住过此宅，故名“皇帝堂”。宅院坐西朝东略偏南，现存墙门、前屋、正屋、后屋等，前后三进，占地面积 570 余平方米。前屋通面阔三间，正中为祖堂，后屋为两层木结构楼房。图为剃头师傅在为屋主理发刮胡子。

Imperial Hall

It is located at No. 18, Liufang Community, Changguo Street, Dinghai District. It is the ancestral home of the local Wang family. It has a history of more than 300 years. It is said that Qianlong lived here when he was in Jiangnan, hence the name “Imperial Hall”. The house sits facing southeast, and many of the old features still exist such as the door wall, front house, main hall and even the back houses.

发蒙学堂

位于舟山市普陀区展茅街道上泮孙村西北角将军路，据《展茅镇志》记载建于清光绪三十二年（1906年），占地面积710平方米，建筑面积388平方米。整座建筑坐西朝东，由正屋、南北厢房组成四合院落。正屋通面阔三间，进深九檩九柱，穿斗式梁架，屋面盖小青瓦。南北厢房通面阔三间，现北厢房只剩一间。发蒙学堂旧址属展茅地区最早的私塾。

Enlightenment School

It's located in the General Road, the northwest corner of Shangpan Sun Village, Zhanmao Street, Putuo District, Zhoushan City. It was built in the 32nd year of the reign of emperor Guangxu of the Qing Dynasty (1906), covering an area of 710 square meters with a construction area of 388 square meters. The whole building is situated in the west facing the east, which is a quadrangle courtyard constructive of the main house, the north and south wing-rooms. The main house is three rooms wide and nine purlin columns deep, in a cross girder frame, which is roofed with green tiles. The north and south wing-rooms is three rooms wide, and there is only one room left in the north wing-room. The old site of Enlightenment School is the earliest private school in Zhanmao area.

刘坤记磁器店

位于管庙区 17 号，始建于清嘉庆年间，因民国年间开过磁器店而闻名。民居规模庞大，布局合理，风格古朴。图中是请师傅来修睡觉使用的棕棚。

Liu Kunji Magnet Store

Located at No. 17 Guanmiao District, Changguo Street, Dinghai District, it was founded in the reign of Emperor Jiaqing in the Qing Dynasty, and was famous for opening a magnet shop during the Republic of China. The building was built simply with a reasonable layout.

周宅

位于定海区环南街道千岛社区五联村西蟹峙中岙 18 号。周氏祖先清代时从小猫山迁居至此，建造了此座具有海岛特色的宅院。西蟹峙岛面积狭小，保存旧式民居不多，周宅三合院已极少见。

Zhou's House

It is located at No.18, Zhong'ao, Xixie Island, Wulian Village, Qiandao Community, Huannan Street, Dinghai District. Zhou's ancestors moved from Little Cat Mountain in the Qing Dynasty and built this house with island features.

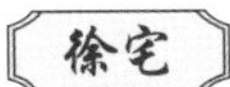

徐宅

位于定海东门右城根 10 号，徐宅有将近百年的历史，规模、格局、面貌基本保存完好。图为徒弟扶着瞎子师傅走街串巷、游走四乡。

Xu's House

It is Located at No. 10 Right City Wall, East Gate of Dinghai. Xu's house has a history of nearly 100 years and the scale, layout and appearance are basically well-preserved. The illustration shows apprentices helping his blind master to cross the street and travel to the village to do some fortune-telling.

施家大院

位于定海区芙蓉弄 5 ~ 9 号，始建于清代，大院坐东北朝西南，占地面积 3030 平方米，建筑面积 1100 平方米。中轴线上有台门、前屋、正屋、后屋等。正屋五间，通面阔 20.4 米，通进深 10.1 米。用七桁，均用穿斗结构，单檐硬山顶。房屋目前保存完好。

Courtyard of Family Shi

It is located in No.5~9, lotus lane, Dinghai District, it was built in the Qing Dynasty. The courtyard faces the southwest and covers an area of 3030 square meters with 1110 square meters of construction area.The house is currently in very good condition.

吴家大院

位于定海区昌国街道前府街 7 号。大院为两层砖木结构，朝南两排，前排有东西厢房，外墙为砖石结构，内部多为木结构。吴氏祖籍宁波慈溪，清朝时奉命来舟山围塘造地，在当地定居建造此屋。现部分吴氏后人迁往海外。

Courtyard of Family Wu

Located at No. 7 Qianfu Street, Changguo Street, Dinghai District. The courtyard is a two-story brick and wood structure with two rows facing south. Wu's ancestral home was in Cixi, Ningbo. During the Qing Dynasty, he was ordered to come to Zhoushanweitang to assert land and finally settled in the local area.

余家大院

位于定海区金塘镇大浦区新道东路 32 号，有一百多年的历史。现台门仍保留原貌，灰塑门楼，雕花精美，正面刻有“新安望族”四字，1950 年改为村办小学，图系当年操办媒事的场景。

Countyard of Family Yu

It is located at No.32 Xindao East Road, Dapu Area, Jintang Town, Dinghai District. The construction has a history of more than 100 years. At present, the Taimen Gate still retains its original appearance. The gray plastic door building is beautifully carved. The front face is engraved with the words "Xin An Wang Zu". In 1950, it was changed to a village-run primary school. The picture shows match-making scenes at that time.

王氏祠堂

位于定海区白泉镇万金湖区繁强村协成一弄。建造历史已逾百年，现存建筑为1920年所建。祠堂坐北朝南为合院式建筑格局，三开间三进，分为门厅、正厅和后厅。建筑面积430余平方米，占地面积600余平方米，是王氏家族用来议事的地方。祠堂建筑不仅在空间上能充分满足族内各种事务的举行，而且建筑精雕细刻，十分考究，有较高的艺术欣赏价值。

Wang's Ancestral Hall

It is located in Xiecheng Fist Lane, Fangxiang Village, Wanjinhu Area, Baiquan Town, Dinghai District. It was built more than a hundred years ago, and the existing structure was built in 1920.The building area is over 430 square meters and the grounds cover an area of more than 600 square meters. It was a place for the Wang family to discuss matters. The architecture of this ancestral hall is exquisite and elegant, and holds a high cultural and artistic value.

虞家东厢门

位于舟山市岱山县岱西镇海丰村虞家山嘴 78 号南侧，单檐硬山顶，台门上方中宫尚残留太极八卦图案，厢门整体保存较完整。虞家东厢门是原虞家庄园的组成部分，旧时的庄园曾经为封闭式，规模较宏大，当地地名即以此来命名。

Dongxiang Gate of Family Yu

It is located at the south side of Yujia Shanzui No.78, Haifeng Village, Daixi Town, Daishan County, Zhoushan City, in a architectural style of gable and hip roof with single eave. The total width of the gate is 3.4 meters, among which the width of the entrance door is 1.6 meters. The pattern of the Tai Ji Eight Diagrams is still left in the Middle Palace above the portal, and the overall preservation of the compartment door is relatively complete. Dongxiang Gate is an integral part of Yujia Manor. In the old days, the manor used to be closed, with a larger scale. The name of the local place is named after it.

蓝府大院

位于定海城区西大街 89 号，始建于清代晚期。蓝府大院坐西北朝东南，占地面积 960 平方米，建筑面积 1317 平方米。中轴线有前屋、正屋、后屋、左右厢房等。正屋五间，通面阔 19.6 米，通进深 10.5 米，重檐硬山顶。蓝氏先人为蓝理（1649 ~ 1719），字义甫，福建漳浦人；康熙二十九年（1690 年）任定海总兵，曾为普陀山修复和舟山的发展做出过重要贡献。

Courtyard of Family Lan

Located at No. 89 West Street, Dinghai District, it was built in the late Qing Dynasty. The Courtyard of Family Lan faces the southeast and covers an area of 960 square meters and with a whole building area of 1317 square meters. Lan's ancestor was Lanli (1649-1719), who served as chief general of Dinghai in the 29th year of emperor Kangxi of the Qing Dynasty .

义桥路35号民居

位于定海区义桥路35号，始建于清代，已有120多年的历史。此宅风格古朴，典型的南方旧式民居。图中居民洗衣晒被，孩子玩耍，门外路边有一走街穿巷卖草鞋的手艺人。

Residence in No.35 Yiqiao Road

Located at No. 35 Yiqiao Road, Dinghai District, it was built in the Qing Dynasty and has a history of more than 120 years. This house is simple in style and typical of old southern houses.

刘鸣生故居

位于定海区东大街聚金弄6号，建于清光绪十四年（1888年）。刘鸣生为定海人，幼年丧父，家道中落，自赴上海谋生，渐在商界露头角，1932年创立华东煤业公司，又组建大中华火柴公司和章华毛纺厂，号称企业大王。图为故居门外编竹笠斗篷的手艺人。

Liu Mingsheng's former Residence

Located at No. 6 Jujin Lane, East Street, Dinghai District, it was built in the 14th year of Emperor Guangxu of the Qing Dynasty (1888). Liu Mingsheng settled in Shanghai and made a modest living. He then gradually emerged in the business world when in 1932 he founded the Huadong Coal Company, the Greater China Match Company and Zhanghua Woolen Mill.

阿斋民居

位于舟山普陀区展茅街道茅洋社区林吴村。当地居民回忆，民国时期，此宅是乔石家中管家吴阿斋的住宅，大四合院，后大部分毁于火灾，现仅存正房、南厢房，上盖青瓦，外墙乱石堆砌。

Azhai's House

Located in Linwu Village, Maoyang Community, Zhanmao Street, Putuo District, Zhoushan. This home once had a courtyard in the shape of a quadrangle until most of it was destroyed by fire. All that remains south wing which is covered in green moss and the outer wall which has stones piled up against it.

刘家老屋

位于岱山县高亭镇沙涂区小蒲门村 102 ~ 105 号。始建于 1915 年，是海岛上典型的南方民居，现极少居民居住，年轻的都外出打工。

Old House of Family Liu

It is located at No.102-105 Small Pumen Village, Shatu Area,Gaoting Town, Daishan County. It was built in 1915 and was built in the typical southern style on the island. Very few residents live there now, and young people go out and work outside.

舟山民居（一）

位于嵊泗县菜园镇青沙村中大街 88 号，建于民国初期，坐北朝南，占地面积 1727 平方米，为前后两个独立三合院组成的建筑整体。是当地保存最为完好的木结构建筑之一，现前院为“青沙渔俗风情馆”，后院为“青沙老年活动室”。

Zhoushan Folk House (1)

Located at No. 88, Middle Street, Qingsha Village, Caiyuan Town, Shengsi County, it was built in the early years of the Republic of China. The home faces the south and covers an area of 1727 square meters and consists of two courtyards, one to the front which is home to the Qingsha fishing pavilion, and the other to the rear which hosts the Qingsha elderly activity room. It is one of the most famous well-preserved wooden structures in the area.

舟山民居（二）

位于岱山县秀山乡秀北村圆墩路 47 号。系清末一高姓大户所建，后屋主因商船失火损失惨重，将此老宅转买。现用作兰秀博物馆展厅，室内陈列秀山文人学士的书画墨宝，是舟山著名的私人收藏博物馆。

Zhoushan Folk House (2)

It is located at No. 47, Yuandun Road, Xiubei Village, Xiushan Township, Daishan County. It was built by the Gao's family in the late Qing dynasty. However, it had to be sold because the owner suffered severe losses on a merchant ship when it caught on fire. Nowadays, it is used as an exhibition hall in the Lanxiu museum.

槐花井

位于定海区小沙镇潭陈村紫微社区侯家旁，因井旁有一棵百年老槐树而得名，当地许多传说都与老槐树有关。槐花井水长年不涸，据传说井水系侯家女儿思念母亲，突水似泉，千年流淌，因而槐树千年不倒。

Flos Sophorae Well

It is situated near Hou's house in Ziwei Community, Tanchen Village, Xiaosha Town, Dinghai District.It is named after a hundred-year-old locust tree near the well. There are many local legends related to the old locust tree which has been standing for a thousand years. The Flos Sophorae well water never dries up all year round. It is said the daughter of family Hou missed her mother so much, the water in the well gushed like a spring without ending from her tears.

六井潭

位于舟山市嵊泗县五龙乡会城岙村幸福弄 18 号，清中晚期为一小井潭，因清澈干甜，水源不枯，又改造成两只并列井潭。是泉城岙最大的井潭，也是当地主要的水源。

Six-well Pond

It's located in the Happy Lane No.18, Huicheng'ao Village, Wulong Township, Shengsi County, Zhoushan City. It was a small pool in middle and later period of Qing Dynasty and then was remodeled and enlarged into two parallel pools because of the endless supply of clear and sweet water. It is the largest pond in Quancheng'ao and the main water source in the area.

自然景观
Landscape

六横岛

位于普陀区，在舟山群岛的南部。元代以前，称为黄公山，明代起开始改名六横岛。因为全岛有东南－西北走向的六条岭横岛屿，而得名“六横”。是舟山群岛中仅次于舟山岛和岱山岛的第三大岛。岛上风景迷人，有沙滩、悬崖、奇石、岩洞、礁滩、原始植被等优美的自然风光。图为六横岛五彩滩。

Six Transverse Island

Located in the Putuo District, in the southern part of the Zhoushan Islands. It was named Huanggong Mountain until Ming Dynasty when its name was changed to Six Transverse Island. The name was changed because the island has six ridges from the south-east to the north-west. It is the third largest island in Zhoushan Archipelago after Zhoushan Island and Daishan Island where the scenery is charming and wonderful.

东福山岛

位于普陀区东极镇，是东海东极列岛的最东端，距大陆岸线最近点约65.4千米，与舟山岛岸距离约42千米，岛大致呈西北－东南走向，长约3千米，宽约1.7千米，面积仅有2.95平方千米。据传因徐福东渡时曾经落脚此岛而得名，被当地渔民戏称为“风的故乡、雾的王国、雨的温床、浪的摇篮”的荒芜之地。但岛上气候独特，风景优美。图为东福山岛船礁。

East Fushan Island

It is located in the Dongji Town of Putuo District. It is the easternmost island in the East China Sea. The island is about 3 kilometers long and 1.7 kilometers wide, it has an area of only 2.95 square kilometers. It is said that Xu Fu had landed on this island during his journey.

枸杞岛

位于嵊泗县菜园镇东 30 千米处。因岛内枸杞岙附近遍生枸杞灌木而得名。是嵊泗列岛中仅次于泗礁山的第二大岛，面积 5.6 平方千米，最高海拔 199 米。岛形略呈“T”字形，以山地为主，山顶多裸岩，沟谷处植被甚茂。

Wolfberry Island

It is located 30 kilometers east of Caiyuan Town, Shengsi County. It is named after the pervasive Chinese wolfberry shrubs near Wolfberry Plain in the vicinity of the island. It is the second largest island in the Shengsi Islands after the Sijiao Mountains. The island is in the shape of a “T” and is dominated by mountains.

五峙山鸟岛（一）

位于舟山本岛西北 7000 米处的五峙山列岛上，是全国三大鸟类保护区之一，五峙山列岛由大五峙山、小五峙山、龙洞山、馒头山、鸦鹊山、无毛山、老鼠山 7 个形态各异的岛屿组成。其中，龙洞山、馒头山、鸦鹊山三个岛屿，每年都会吸引 42 个种类近万只水鸟到此停歇、栖息和繁殖。

Wuzhi Mountain Bird Island (1)

It is located on Wuzhi Mountain Islands, 7 kilometers northwest of Zhoushan Island, it is one of the three major bird sanctuaries in China. It is one of the important breeding sites of wetland waterbirds found around the coast of Zhejiang Province, and plays a very important role in their protection.

五峙山鸟岛（二）

五峙山鸟岛是中国三大鸟类保护区之一，也是浙江省唯一的省级海洋鸟类自然保护区，为浙江沿海一带发现的湿地水鸟重要繁殖地之一。每年有夏候鸟 19 种，冬候鸟 17 种，留鸟 6 种到此繁衍栖息。五峙山列岛鸟类自然保护区在湿地水鸟保护中有着非常重要的作用。

Wuzhi Mountain Bird Island (2)

Wuzhishan Bird Island is one of China's three major bird sanctuary and the only provincial marine birds Nature Reserve in Zhejiang Province, made of the coastal areas of Zhenjiang. It is now one of the important wetland waterfowl breeding grounds. The Wuzhi Mountain Islands Birds Nature Reserve has a very important role in wetland waterfowl conservation.

沈家门渔港

位于舟山本岛东南侧，面临东海，背靠青龙、白虎两山，是中国最大的天然渔港，与挪威的卑尔根港、秘鲁的卡亚俄港并称世界三大渔港。这里早在清代便形成了热闹的街市，曾有“市肆骈列，海物错杂，贩客麇至”的记载。每逢渔汛，沿海十几个省市的几十万渔民云集港内，桅樯林立，鱼山虾海，形成了一道独特的海岛渔港景观。

Shenjiamen Fishing Port

It is located on the southeast side of Zhoushan Island, facing the East China Sea, backed by the mountains of Qinglong and Baihu. It is the largest natural fishing port in China which is known as one of the three largest fishing ports in the world together with Bergan Port of Norway and the Callao Port of Peru. During the fishery season, fishermen gather in the harbor, with numerous mountains of fish and shrimp, forming a unique harbor landscape.

双屿港

位于普陀区六横岛中间，16 世纪时是当时亚洲最大的海上走私贸易基地，双屿港悬居海洋之中，处于主航道线上，但又距海岸不远，便于粮草接济，正是从事走私贸易的好地方。当时日本、葡萄牙等地的贸易客多以此地为据点，从事商业走私活动。

Double Islands Port

It is located in the middle of Six Transverse Island in Putuo District. It was the largest maritime smuggling trade base in Asia in the 16th century. Double Islands Port lies in the ocean and on the main navigation channel, near the coast so that it was convenient for food supply. Most traders from countries such as Japan, Portugal and others used this place as their base to engage in their smuggling activities.

定海道头

位于定海区沿港东路与和平路交叉路口处。明代称"舟山渡"，清代称"定海山渡"，后又名"定海港"。始建于宋代，是舟山群岛历史最久远、船只进出最频繁、使用率最高的往来港口，还是浙东沿海最早与外国通商的港口。

Dinghai Port

Located at the intersection of Yangang East Harbour Road and Heping (peace) Road in Dinghai District. It was founded in the Song Dynasty, called "Zhoushan Ferry Crossing" during the Ming Dynasty, and then changed to "Dinghai port" in the Qing Dynasty. It is the oldest and busiest port in Zhoushan district with the highest number of ships entering and leaving the district, which started when trade began with foreign countries along the eastern coast of Zhejiang.

大鹏渡口

位于定海区金塘镇大观社区大鹏岛。是一个清代所建的老码头，对岸为金塘沥港，大鹏渡口是旧时大鹏岛与金塘来往的必经渡口，码头由块石垒砌，长约 60 米。20 世纪 70 年代后逐渐废弃。

Dapeng Ferry Crossing

Located on Dapeng Island, Daguan Community, Jintang Town. It is an old wharf built during the Qing Dynasty. The Dapeng Ferry Crossing was the only means for reaching Jintang, which was the destination on the other side of the crossing. The Dapeng Ferry Crossing was gradually abandoned after the 1970's.

短姑道头

位于普陀区普陀山风景区的南入口处。又称短姑圣迹、短姑古迹。此地旧时原为海滩，滩上有阔十余米，长百来米，两侧错列大小不一、形状各异的岩石，岩石上镌有“短姑古迹”“佛放光明”“同登觉岸”“乐土”等14款题刻，出没于潮汐浪涛之中。

Duangu Harbor

It is located on the southern entrance of Putuo Mountain Scenic Area in Putuo District. Also known as short sacred traces, short sacred monuments or Duangu Miracle. The old place was originally a rocky beach of more than ten meters wide and one hundred metres long. There are small monuments and shrines to Buddha on the rocks, and sometimes when the tide is out, you can find 14 inscriptions written on some of the rocks, which include “shore”, “happy land” and so on.

千步沙

位于普陀区普陀山东部海岸。总长近2000米，因其长度近千步而得名。是普陀山上最大的沙滩，沙质细腻。沙滩沙坡平缓，海面宽阔，且水中无乱石暗礁，是踏浪、观海、游泳的好地方。

Thousand-Step Sand

Located on the eastern coast of Putuo Mountain. The total length is nearly 2 kilometers, and it is named after its length of nearly a thousand steps. It is the largest beach on Mount Putuo and has smooth fine sand. It is considered a fine place to go for a walk on the beach, to go swimming and to just sit and watch the waves of the sea.

海岸牌坊

位于普陀区普陀山风景区的南面，始建于1919年。四柱三门，翠瓦飞檐，上有“南海圣境”“同登彼岸”“宝伐迷津”“金绳觉路”“回头是岸”匾额五重，为北洋政府黎元洪、徐世昌、冯国璋等人所题。

Coastal Archway

Located in the south of Putuo Mountain Scenic Area in Putuo District. It was founded in 1919. It consists of four pillars and three gates, and is decorated with black tiles and cornices. There are five plaques on the archway which have been inscribed by Li Yuanhong, Feng Gouzhang and others from the Beiyang government.

南天门

位于普陀区普陀山南的南山上，与短姑道头对峙。南天门孤悬入海，处于普陀山最南端，与本岛一水相隔。此地巨石森立，危岩高耸，中有两石如门，故名南天门。阙门飞檐起角，中间书有“南天门”三字，旁有龙眼井，崖上有石鼓，阙左上方有狮子石。

The Heavenly Southern Gate

It is located on the south area of Putuo Mountain in Putuo District, facing Duangu harbour. The heavenly Southern Gate is separated from Putuo Island by the sea, it is a lonesome figure which extends into the sea as the southernmost point of Putuo mountain. There are numerous huge stones and high rockfaces of which two stands apart from each other looking like a gateway to the island, hence the name, heavenly southern gate.

说法台

位于普陀区普陀山西南岸，磐陀石附近。相传为观音菩萨当初说法时的法座，四周散落着许多岩石，或大或小、形状各异，或三三两两聚在一起，像在聚精会神听法。此处的摩崖石刻很著名，图为“说法台”“法台灵迹”石刻。

Buddhist Pulpit

Located on the southwestern coast of Putuo Mountain, near the Pantuo Stone. According to legend, it was the seat of Guanyin Bodhisattva. There are many rocks of all shapes and sizes, scattered around the area in two or three, as though they are listening to sermons. There also stands some very famous cliff stone carvings. The picture shows the stone carvings of Buddhist Pulpit and the spiritual manifestation.

朝阳洞

位于普陀区普陀山镇前山村珠宝岭观音阁东下崖间。因每日太阳东升时，阳光倒映此洞，而得名“朝阳洞”。朝阳洞外巨石参差，洞口直面东海，左右挽百步沙与千步沙。在普陀山观日出，以朝阳洞为最，故有“朝阳涌日”之说。

Greeting Sun Cave

It's located in the Dongxia Cliff of Guanyin Pavilion, Jewelry Hill, Qianshan Village, Putuo Mountain Town. The cave is called Greeting Sun Cave because the sunshine falls on the cave when the sun rises eastward every day. There are uneven huge rocks outside the cave, and the entrance to the cave directly face the East China Sea. It's the best place to watch the sunrise in Putuo Mountain.

白龙潭

位于定海区岑港街道司前社区。潭最深处 16 米，潭口直径宽 3 米，潭中间还有一把沙发样的石头，可以睡觉的，很光滑。民间传说白龙潭曾为白老龙修炼处，为获蟾成仙，白老龙曾化渔民帮助寡妇捕鱼，得蟾成仙后，逢旱降水，造福岑港一带百姓。

White Dragon Lake

It is located in the Siqian Community, Cengang Street, Dinghai District. The depth of the pond is 16 meters and the width is 3 meters. There is a smooth sofa-like stone in the middle of the pond. Legend says it is the place where an old white dragon became immortal through practice.

倚剑摩崖石刻

位于嵊泗县大洋镇小洋村小观音山朝西南山麓中，石刻面朝西，分布面积 27 平方米，横书楷体阳刻，每字高 3.8 米，宽 3.5 米，左落款为“癸丑夏楚人李楷书”。据落款及史籍记载，为嘉靖三十二年（1553 年）举人昌乐县知县李楷所书，倚剑二字堪称“大字王”，遒劲奔放，豪气撼人，是舟山市迄今发现的最大的摩崖石刻。

Leaning Sword Cliff Inscription

It's located in the southwestern foothill of Small Guanyin Mountain, Xiaoyang Village, Dayang Town, Shengsi County. The incised inscription faces the west and covers an area of 27 square meters, in horizontal mode, regular script and garland. According to the inscribes and historical records, it was written by Li Kai, magistrate of Changle County, in the 32nd year of Jiajing of the Ming Dynasty (1553). It's the largest cliff inscription ever found in Zhoushan City.

小观音山古摩崖石刻

位于嵊泗县洋山镇观音山顶，共有古摩崖石刻八处，最早的一方题刻“海阔天空”镌于明万历三十六年（1608 年）。其他题刻如“鲲鹏化处”“中流砥柱”“水天阔处”和“海晏波宁”等，皆浑雄大气，据考证这些石刻均系明代所题。这里千米山路中危崖夹峙，奇石相随，也是明代抗倭遗址。

Ancient Cliff Inscription in Small Guanyin Mountain

It's located at the lop of Cuanyin Mountain, Yangshan Town, Shengsi County. There are eight ancient cliff inscriptions, with the earliest inscribed in the 36th year of Wanli in the Ming Dynasty (1608). It's proved that these inscriptions were all inscribed in the Ming Dynasty. Here steep cliffs tower around kilometres of mountain roads with rugged stones. It is the relics of the battlefield resisting Janpanese pirates in Ming Dynasty.

百步沙摩崖石刻

位于普陀区普陀山镇前山村百步沙沙滩的巨崖上，镌面朝西，背对大海，刻有“师石”“回头是岸”两款。图为用篆书书写的“师石”款，字迹秀丽，刚劲端庄，字义与环境相映成趣。

Baibusha Cliff Inscription

It's located on the huge cliff of Baibusha Beach in Qianshan village, Putuo Mountain Town. Facing the west and backed by the sea, the cliff is inscribed with "Master Stone" and "Repent and Be Saved". The handwriting is beautiful, vigorous and dignified, with the meaningful words in harmony with the environment.

磐陀石摩崖石刻

位于普陀区普陀山梅岭峰梅福庵西侧，一片开阔的山顶平台上。相传金地藏入九华曾在此岩石上打坐禅修，后人称此“打坐石”又称“磐陀石”。今石壁上有“金地藏第一修行处”的摩崖石刻。

Pantuo Cliff Inscription

It's located on the west side of Meifu Nunnery in Meiling Peak, Putuo Mountain. It's a flat rock on the top of mountain. It's said that Jin Dizang, a monk who went to Jiuhua Mountain sat in meditation on the rock. The rock now is called "Meditation Rock" or "Pantuo Rock". Nowadays, the cliff is inscribed with " the first place where Jin Dizang meditated".

嵊泗“山海奇观”摩崖石刻

位于嵊泗县枸杞乡里西村里西岗墩峰顶巨石上，刻于明万历十八年（1590年）。石东壁面朝东方镌刻“山海奇观”四个大字，下有落款：“大明万历庚寅春，都督侯继高统率临观把总陈九思、听用守备宋大斌、游哨把总詹斌、陈梦斗等督汛于此”，该石刻刻工较为精良，为研究明代抗倭历史提供了实物依据。

Shengsi Cliff Inscription of “Mountain and Sea Wonders”

It's situated on a huge stone, at the top of Lixi Mound Peak, Lixi Village, Wolfberry Village, Shengsi County. It was inscribed in the 18th year of Wanli in the Ming Dynasty (1590). The East Stone Wall faces the east and is engraved with four huge delicately-carved characters “Shan Hai Qi Guan”, providing a physical basis for the studies of the anti-Japanese battle in the Ming Dynasty.

东崖绝壁

位于嵊泗县菜园镇正东35千米处。东崖绝壁既是嵊山岛的最东端，也是舟山群岛的最东端。绝壁的岩性为浅肉红色钾长花岗石，北自后岭头屿，南至鳗鱼头岛，险峰连绵高耸。山势自山腰颓然直泻入海，形成高崖绝壁，上耸数十丈，绵延3000米，最高处达70余米。沿岸景致奇幻壮丽，动人心魄。

East Cliff

It's located 35 kilometers to the east of Caiyuan Town, Shengsi County. East Cliff is the most eastern point of the Zhoushan Archipelago. The precipice falls into the sea from the mountainside, forming a high cliff soaring to the sky, which stretches for 3 kilometers and reaches the highest point of over 70 meters. The scenery along the coast is fantastic, magnificent and striking.

两龟听法

舟山群岛普陀山林木葱茏，洞壑幽深，奇石嶙峋，潮音不断。岛上岩石因海风蚀化而千奇百怪，姿态万千。二龟听法位于普陀区普陀山梅岭峰梅福庵西侧的岩崖上，两石酷似海龟，一龟蹲踞崖顶，回首顾盼，似有等候之意。另一龟缘石直上，昂首延颈，筋膜尽露，一副着急相。两龟的形态极为传神。

Two Turtles Listening to Sermons

Located on the cliff to the west of Meifu Nunnery, Meiling Peak, Putuo Mountains, two rocks resemble two sea turtles. One turtle crouches on the cliff top, looking back as if it was wating for someone. The other turtle is straight up, head raised and neck stretched, fascia exposed, with an anxious look. The shape of both tortoises is rather vivid.

静室茶烟

“静室茶烟”是明代“普陀十二景”中的一景。在佛顶山慧济寺后（当地俗称后山），曾有一片古茶山，在古茶山一带有 17 处庵院，静室即指这 17 处庵院，而茶烟则特指煮水泡茶时产生的烟。普陀山最早的茶叶种植始于南宋年间，而明万历年间和清康熙年间，是普陀山茶叶最兴旺时期，当地的云雾佛茶以其“叶厚芽嫩”“叶有白茸”而闻名。

Quiet Chamber and Tea Steam

"Quiet Chamber and tea steam" is one of the twelve scenes of Putuo in Ming Dynasty. Behind Huiji Temple in Foding Mountain (commonly known as Houshan), there was an ancient tea mountain. In the tea mountain, there were once 17 nunneries which are what the quiet chambers refer to, while the tea steam is the steam produced when tea is boiled and brewed.

马岙盐场

位于定海区马岙镇北海村庙山西北侧山麓，占地 1400 多亩。此盐场现今已不再使用，但曾经规模很大，是舟山晒盐业发展代表性的物证，有较高的历史研究价值。

Ma'ao Saltern

Located in the foothills of Beihai Village to the northwest of MiaoShan Mountain, Ma'ao Town, Dinghai District, the saltern covers an area of more than 1400 mu. Though it is no longer in use, it provides significant evidence for the development of sun-drying salt industry in Zhoushan due to its large scale and historical value.

双合石壁（一）

位于岱山县岱西镇双合村，此地的石板因质细而坚韧享有盛名，有五六百年的开采历史。所采石料分石板、石条、石块等，当年江浙地区的建筑所用石材大部分出自此处。因开凿留下的痕迹，至今留下石景旧迹50多处，奇石怪洞，形各有状。

Shuanghe Cliff (1)

Located in Shuanghe Village, Daishan County, the stone slabs there are famous for their fine and tough quality and enjoy a mining history of five or six hundred years. The quarries are divided into slabs, strips, stones and so on. Most of the stones used in the buildings in Zhejiang region at that time came from the village.

双合石壁（二）

舟山的采石业真正形成规模，是在民国年间，大约有四五十宕，还成立了公司。他们开宕口、打石头就地取材打制石板、石条等重要的建筑材料。除了能满足当地海岛老百姓日常所需外，还远销到上海等地。

Shuanghe Cliff (2)

The scale of Zhoushan's quarry industry came into being was during the era of the Republic of China, with about forty or fifty workshops and even companies. The produced building materials could not only meet the daily needs of local people, but also were exported to Shanghai and other places.

生活民俗
Folk Life

渔场生活（一）

舟山渔场地处东海，是中国最大的近海渔场，自古以来因渔业资源丰富而闻名，是浙江省、江苏省、福建省和上海市三省一市渔民的传统作业区域。自开发以来，一直为沿海渔民共同捕捞场所。旺汛高峰时渔船上万艘、作业渔民在 15 万人以上。

Fishery Life (1)

Zhoushan Fishing Ground is the largest offshore fishing ground in China and it has been famous for its rich fishery resources since ancient times. It is a traditional work area in Zhejiang, Jiangsu, Fujian and Shanghai.

渔场生活（二）

舟山群岛因海而生，因水而兴，城垣是古城的形体，水系是古城的血脉，文化是古城的灵魂。经济与发展不断地循环，周而复始，渔汛到来，千帆竞发。

Fishery Life (2)

Zhoushan Archipelago was born by the sea and thrived by the water. The city wall is the shape of the ancient city, and the water system is the artery.When the fishing season comes, thousands of boats sail.

渔场生活（三）

舟山渔场是众多的经济鱼虾类的产卵、索饵场所，以大黄鱼、小黄鱼、带鱼和墨鱼（乌贼）四大家鱼为主要渔产。同时也是中国规模最大、产量最多的带鱼渔场，其良好的海底地势是拖网作业的良好区域，因此也成为了全国最著名的带鱼捕捞地。图为捕带鱼归来后码头的繁忙景象。

Fishery Life (3)

Zhoushan Fishing Ground is a spawning and baiting place for many economic fishes and shrimps. The main fisheries are big yellow croaker, small yellow croaker, hairtail and cuttlefish (squid).It is also China's largest and most productive hairtail fishery.The picture shows the busy wharf upon the return of hairtail fishing.

渔场生活（四）

舟山群岛每当渔汛季节，岛上除老人、妇女小孩外，男壮力都出海捕鱼一去多日。岛上妇女日夜不歇的织网补网。

Fishery Life (4)

When it is fishing season on Zhoushan Archipelago, except for the elderly, women and children, men go fishing for days.Women usually engage themselves in net weaving day and night.

渔场生活（五）

渔业的发达促使造船业也蓬勃发展，这是当地渔民在修造木帆船，用以在渔汛季节到来时出海捕鱼。

Fishery Life (5)

With the development of fisheries, the shipbuilding industry is also booming. This picture shows the building of wooden sailboats of locals before the fishing season comes.

渔场生活（六）

渔民在大海即将涨潮时出海，涨潮时便开始在绑着网绳的浮标之处收渔网。随着长长的渔网一段段被收上渔船舱，又大又长的捕鱼专用网兜露出了水面，网兜里面活蹦乱跳的鱼虾被收进舱内，渔民挑拣完鱼虾后，又将渔网翻开，重新放入海中，借助潮水的冲力，渔网自动展开，为下一次捕鱼做好了铺垫。大海退潮，渔民满载而归。

Fishery Life (6)

Before the sea rises, fishermen go to the sea. They collect the fishing net at high tide, then the live fishes and shrimps in the net are collected into the cabin. After the fishermen have picked up the fishes and shrimps, they turn the net over and put it back into the sea. With the help of the waves of the tide, the fishing net is automatically open, which prepares for the next fishing.

渔场生活（七）

男人们出海，一去就是几周，甚至几月，过去通信不便，只得在海边盼归来帆影，牵肠挂肚，日思夜想，度日似年……

Fishery Life (7)

Men go to sea for weeks or even months. In the past, for the lack of communication, women could not reach them. They had to wait by the sea, looking forward to the return of the sails, worrying and thinking about their men day and night. One day seems to be as long as a year.

拾海螺

舟山群岛中，嵊泗、嵊泗因海岛纷杂，渔产品甚丰，海滩海涂上遍产海螺、泥螺，大人出海捕鱼，孩子们则在附近海滩海涂中捡拾海螺、泥螺等海产品。

Collecting Conch

Out of all Zhoushan Archipelago, Shengshan and Shengsi are rich in fishery products such as conch and mud snail, because of their various islands. When adults go fishing, children collect conches, mud snails and other seafood in the nearby beach.

晒鱼干（一）

这是普陀区普陀山虾峙岛黄石渔村渔民们在晒鱼干，因新鲜鱼类过去没有冰箱或其他制冷工具用来保鲜，捕上来的鱼类几天后即腐烂生虫，只有晒成鱼干方能保存长久，待需要时食用。

Basking Fish (1)

Fishermen in Huangshi Fishing Village of Xiazhi Island near Putuo Mountain, are drying fish. There were no refrigerators or other icing devices in the past, so freshly-caught fishes were easily rotted just in a few days. Only dried fish can be kept for a long time and then be eaten when needed.

晒鱼干（二）

舟山群岛每当渔汛时节，大小黄鱼纷纷登场，渔民们从头至尾一刀剖开，用盐腌制，压榨十几天后，取出爆晒二三天后上篓，运往内地销售，八九月间运到温州销售，每季可达几万条之多。

Basking Fish (2)

When it is fishing season on Zhoushan Archipelago, yellow croakers with different sizes can be seen everywhere. Fishermen cut them open from head to tail, marinate them with salt, press them for more than 10 days, bask them for a couple of day and then pack into baskets to ship them to the mainland of the Yangtze River for sale. In August and September, they are shipped to Wenzhou for sale. The number of fishes can be up to several tens of thousands each season.

舟山龙船

每当五月端午佳节，舟山各地方乡村社团，自发组织龙船比赛，观者人山人海，呼声雷动，自是一番欣欣向荣的繁华景象。

Zhoushan Dragon Boat

Every May when it is Dragon Boat Festival, local village associations in Zhoushan spontaneously organize Dragon Boat races. The crowd of visitors makes thunderous voice, which is like a thriving and prosperous picture.

舟山舞龙

舟山舞龙俗称“调龙”和“滚龙灯”。龙灯大多由竹木、彩纸、布等扎成，龙身由九节或十一节组成，龙头用竹条扎成架子，糊上白色清明纸涂上各种颜色，有角、嘴、眼、胡须等，形态逼真。龙身各节内能燃烛的称“龙灯”，不能燃烛的称“布龙”。为了祈求风调雨顺，舟山地区的各个海岛每年会在元宵节前后举行盛大的舞龙灯活动。

Zhoushan Dragon Dancing

Zhoushan Dragon Dancing is commonly known as “Tiaolong” and “Rolling Dragon Lantern”. Dragon lanterns are mostly made of bamboo, paper, cloth, etc. They are bright in color and lifelike in shape. In order to pray for good weather, every island in Zhoushan will hold a grand dragon lantern dance around the Lantern Festival every year.

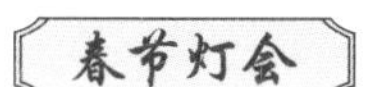

正月里家家户户挂红灯，每逢过年过节，新春来临时，能工巧匠就各显奇能，置灯结彩。特别每年正月十五元宵佳节，舟山的各个海岛上更是到处灯火辉煌，热闹非凡，庙会灯会层出无穷。

Spring Festival Lantern Festival

Every household hangs red lights when it is the first month in Chinese lunar calendar. In festival season, Spring Festival in particular, skilled craftsmen display their unique abilities to make lanterns with different colors. On the Lantern Festival, which is the 15th day of the first month in Chinese lunar calendar, every island in Zhoushan is full of brilliant lights and extraordinary liveliness, and there are endless lamp fairs at Temple fairs.

男婴满月海边织水

舟山群岛民风历来古朴淳厚，该地有一民俗，即男婴长至满月，放入一尺大脚桶，带到海边，让他与海水交流，听海涛，吹海风，使他自小便与海交融在一起。

Baby Boys Learn about Water by the Sea in Their First Month of Life

Zhoushan Archipelago has always been of simple and honest folk customs. There is a custom that people put their baby boys, who are in the first mouth of life, into a one-foot-size bowl and go to the seaside so that the babies can communicate with the sea. Listening to waves, feeling the breezes, the babys have been connected with the sea since young age.

奉鸡成亲拜堂

舟山地区一直保留着古老的用公鸡代新郎成婚的仪式。特别是渔民出海一去几个月，甚至几年不回来，只好请媒婆找一女子，选好吉日，用公鸡代新郎拜堂成亲，表示吉祥如意，“吉”即“鸡”。

Perform a Formal Wedding Ceremony by Using a Cock to Replace the Groom

Zhoushan has preserved the ancient wedding ceremony of marrying a cock instead of a groom. When fishermen go to the sea, they will not return until months or even years later. Some family have to find a matchmaker to find a woman, choose an auspicious date, and use a cock to represent the groom to complete the wedding ceremony. The pronunciation of cock is almost the same as that of auspice in Chinese, which explains why cocks represnt grooms.

布袋木偶戏

定海区双桥镇的布袋木偶戏（俗称小戏文）源于明代。其最大特点是能一人操纵和演出多个人物角色，不像其他木偶戏只能是“一人一偶”地表演。同时，由于没有繁复的道具和装备，可以“轻车简从”地挑着一副担子走巷穿户地进行表演，也称为“扁担戏”。目前双桥镇仍有 18 个木偶戏班，40 多名表演艺人，被称为“布袋木偶戏艺术之乡”。

The Puppet Show

The Puppet Show (commonly known as Xiaoxi Wen) in Shuangqiao Town, Dinghai District originated from the Ming Dynasty. Its greatest feature is that one person can manipulate and perform multiple characters, without complicated props and equipment. They can perform from lane to lane, traveling light. It's also known as "pole play".

舟山锣鼓

以锣、鼓、钹及锁呐为基调，间以丝竹，旋律紧凑、气氛热烈，具有鲜明的海岛特色。明清时期开始广为流传，最早起源于航海，早先在码头用锣鼓招徕乘客及途中娱乐和海上遇险时作求救信号。后来逐渐被吸纳进日常生活中，婚嫁喜庆、祝寿做生，新船下海、乔迁新居、开张营业，都少不了约请一班鼓手吹打一番。

Zhoushan Gong and Drum

With gongs, drums, cymbals and locks providing the keynote, and silk bamboo sound mixed in, the melody is compact and the atmosphere is warm, with distinctive island characteristics. Local people always invite a group of drummers to play when celebrate their marriage and birthday, launch a new ship, relocate and open a new business.

舟山白鹅

又称“浙东白鹅”，其外形美观，羽毛洁白，体重四五千克，定海区的白泉镇和马岙镇，因盛产此鹅而被称为“浙江白鹅之乡”。

Zhoushan White Goose

Also known as “Zhejiang Eastern White Goose”. They have beautiful appearances and white feathers, with weight of four to five kilograms. Baiquan Town and Ma’ao Town in Dinghai District is known as “Zhejiang White Goose Town” because of their abundant geese.